Book of Venus for you

A BOOK OF
VENUS
FOR YOU

MARINER II ON TOP OF
ATLAS-AGENA B MISSILE

A BOOK OF

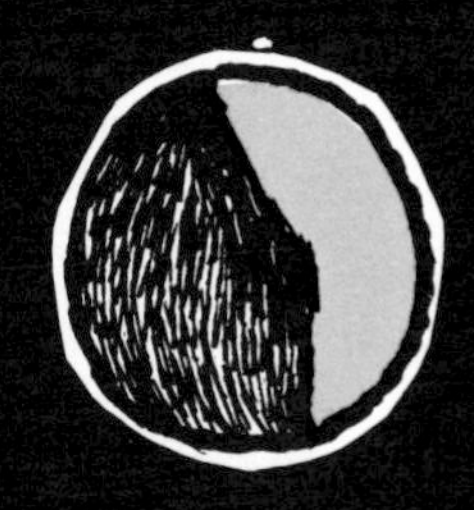

VENUS FOR YOU

BY FRANKLYN M. BRANLEY

Illustrated by Leonard Kessler

Thomas Y. Crowell Company New York

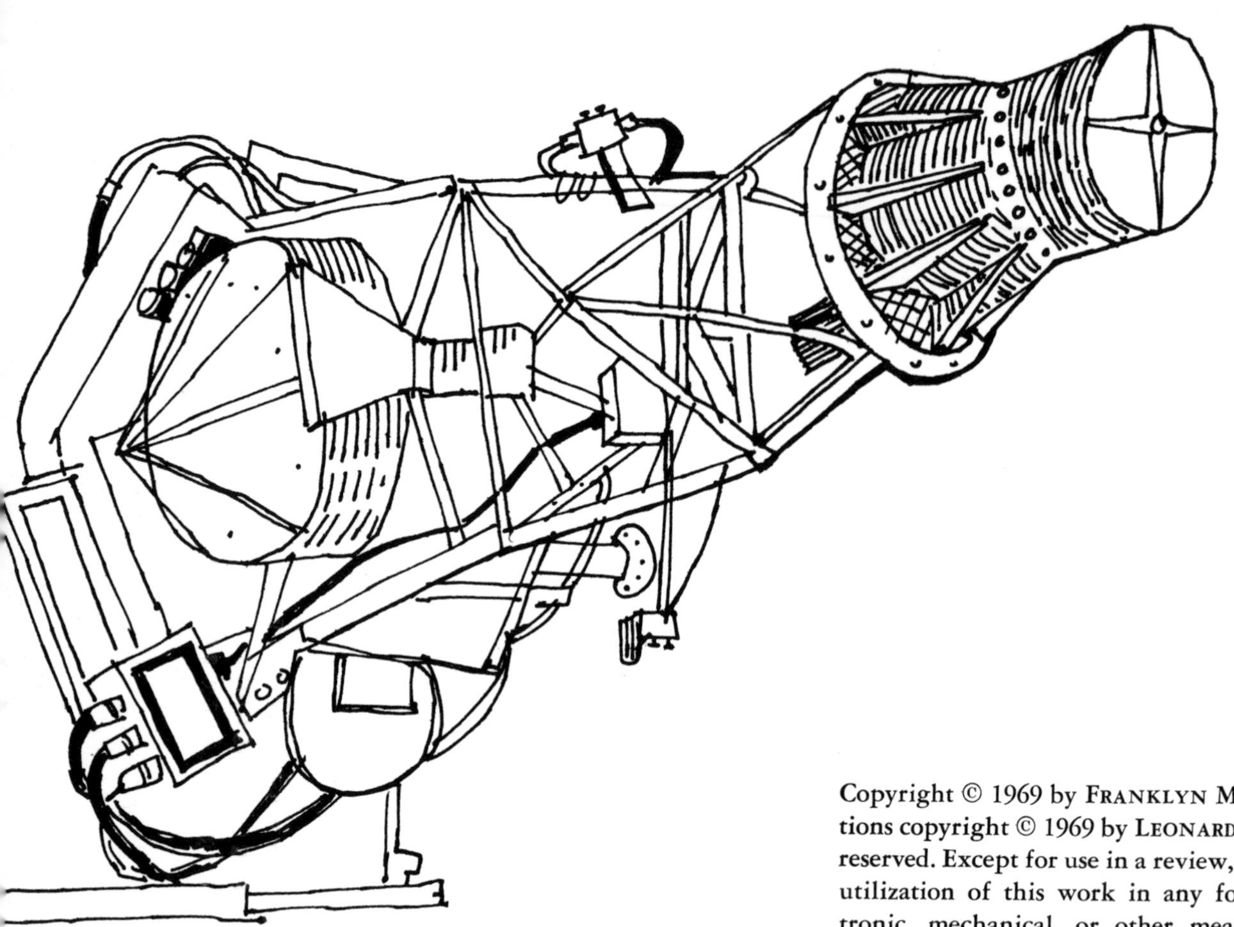

Manufactured in the United States of America.

L.C. Card 73-78256

1 2 3 4 5 6 7 8 9 10

BY THE AUTHOR

A Book of Satellites for You
A Book of Planets for You
A Book of Astronauts for You
A Book of Moon Rockets for You
A Book of the Milky Way Galaxy for You
A Book of Stars for You
A Book of Mars for You
A Book of Venus for You

To K. E. D.

Men have looked at Venus for thousands of years, but no one has ever really seen the planet. That is because Venus is surrounded by clouds through which we cannot see. The clouds reflect the sunlight that falls on them just as a mirror reflects light. When we see Venus shining in the sky, we see only reflected sunlight from its topmost layer of clouds.

Although no one has ever seen what is beneath the clouds of Venus, we have learned a lot about the planet.

Early men saw Venus in the evening and called it an evening "star." They didn't know that planets produce no light of their own, as a star does. They didn't know that planets move around the sun in orbits. Early men thought that everything they saw shining in the sky was some kind of star. There were "shooting" stars (the meteors); there were "long-haired" stars (the comets); and there were the "wandering" stars (the planets). The planets were called wandering stars because they change their positions in the sky from day to day as they move through their orbits. If a planet was seen after sunset it was called an evening star.

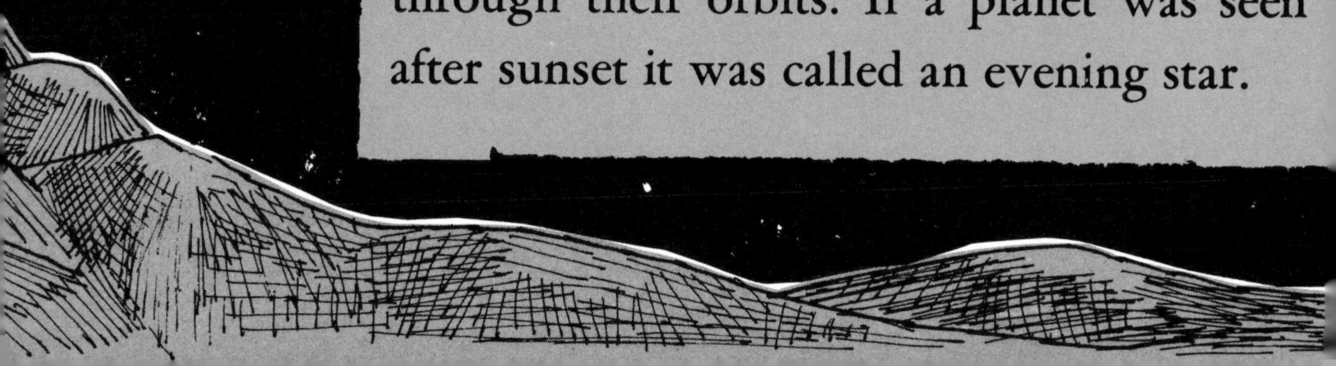

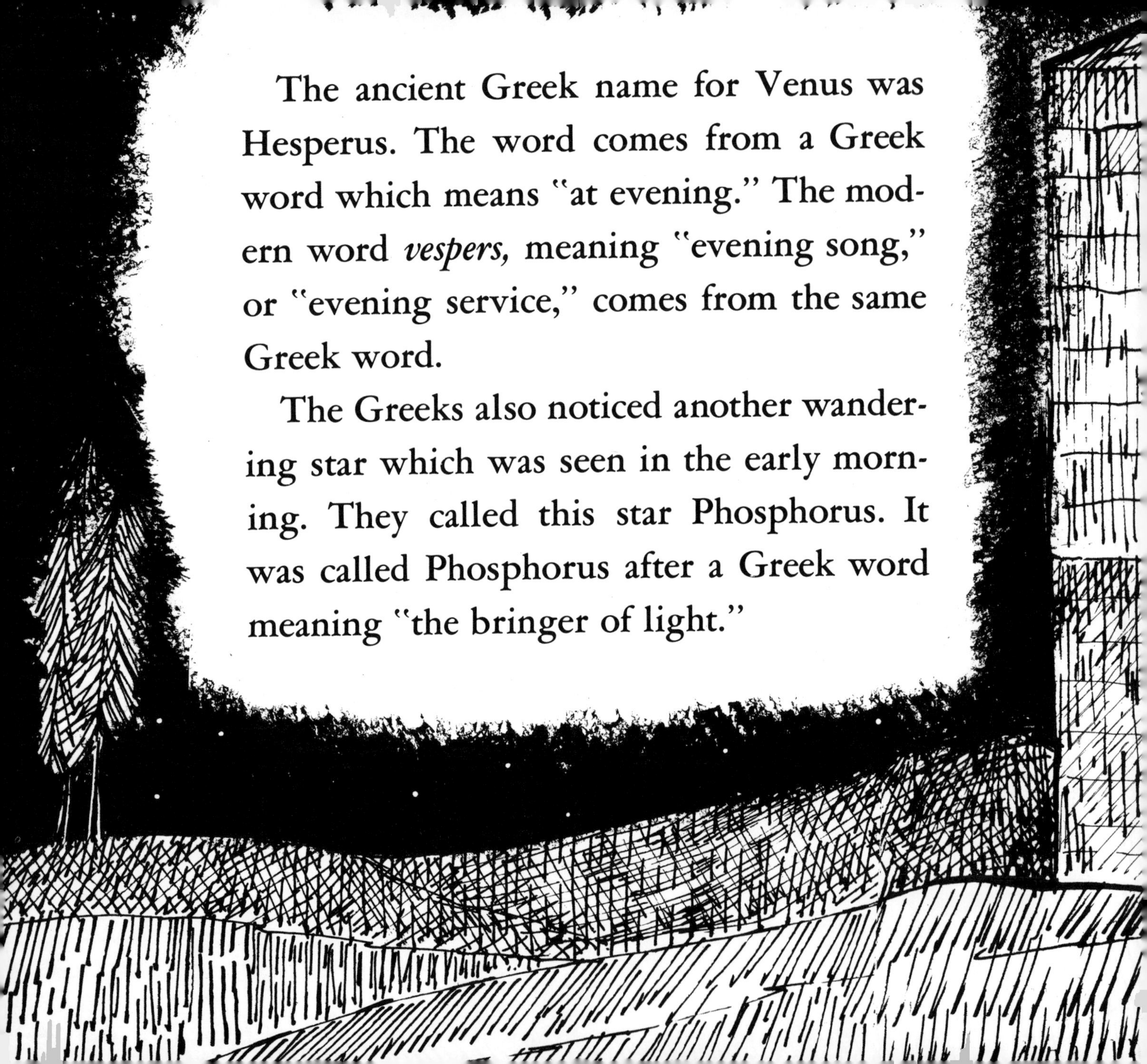

The ancient Greek name for Venus was Hesperus. The word comes from a Greek word which means "at evening." The modern word *vespers,* meaning "evening song," or "evening service," comes from the same Greek word.

The Greeks also noticed another wandering star which was seen in the early morning. They called this star Phosphorus. It was called Phosphorus after a Greek word meaning "the bringer of light."

But not all the Greeks believed that Hesperus and Phosphorus were two different stars. About 2,500 years ago a Greek philosopher named Pythagoras figured out that the two stars were really the same one. Pythagoras had no telescope to help him view the planets. To this day we do not know how he made his discovery. Perhaps he figured it out because he noticed that Hesperus and Phosphorus were never in the sky on the same night.

Today we know why Venus is sometimes an evening star and sometimes a morning star and why we can never see Venus in the middle of the night.

The orbit of Venus is nearer to the sun than the orbit of the earth is. At midnight the side of the earth we are on is facing away from the sun. We cannot see the sun, and we cannot see Venus. The planet Mercury is even closer to the sun than Venus. We can't see Mercury in the sky at midnight, either. Venus and Mercury are always on the same side of the earth as the sun is.

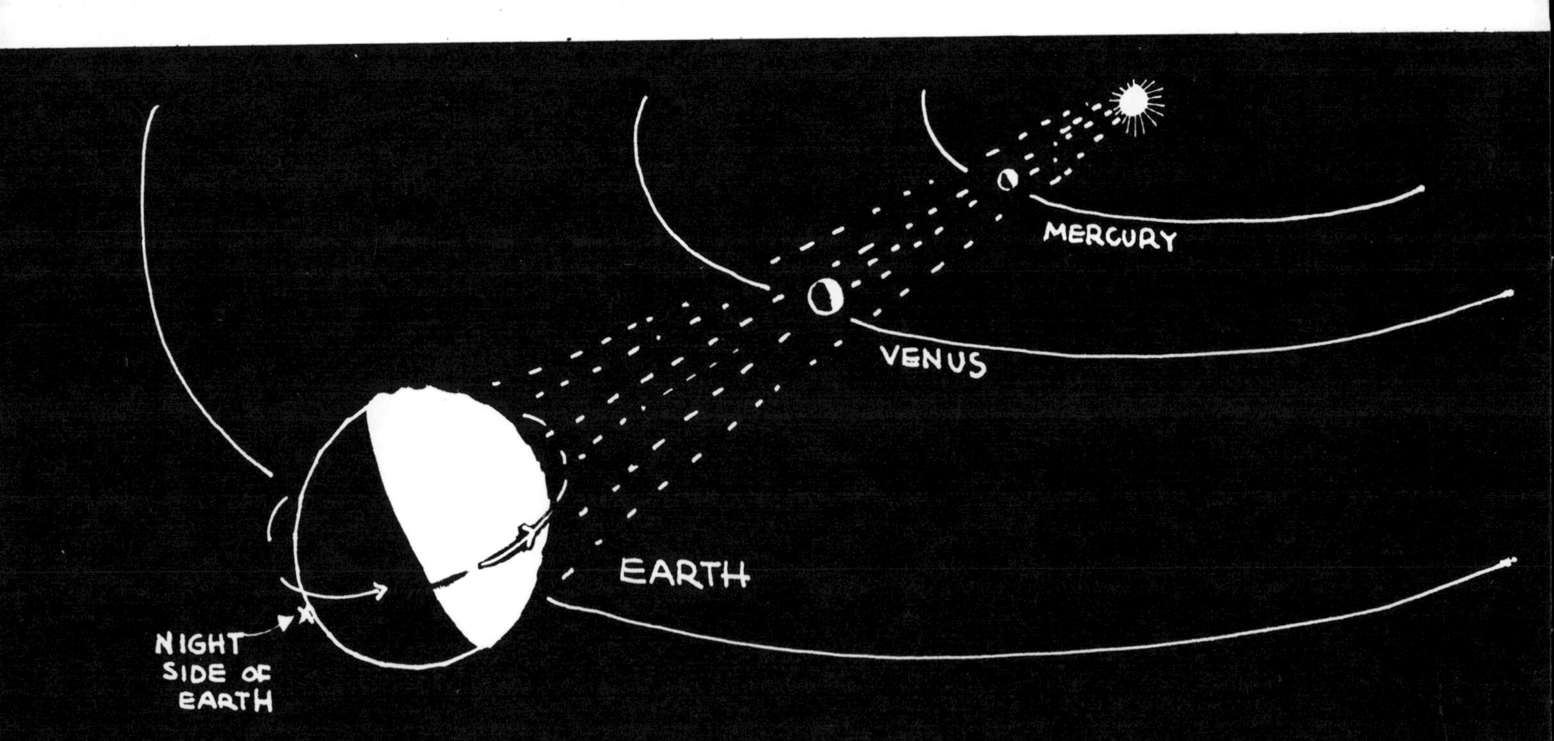

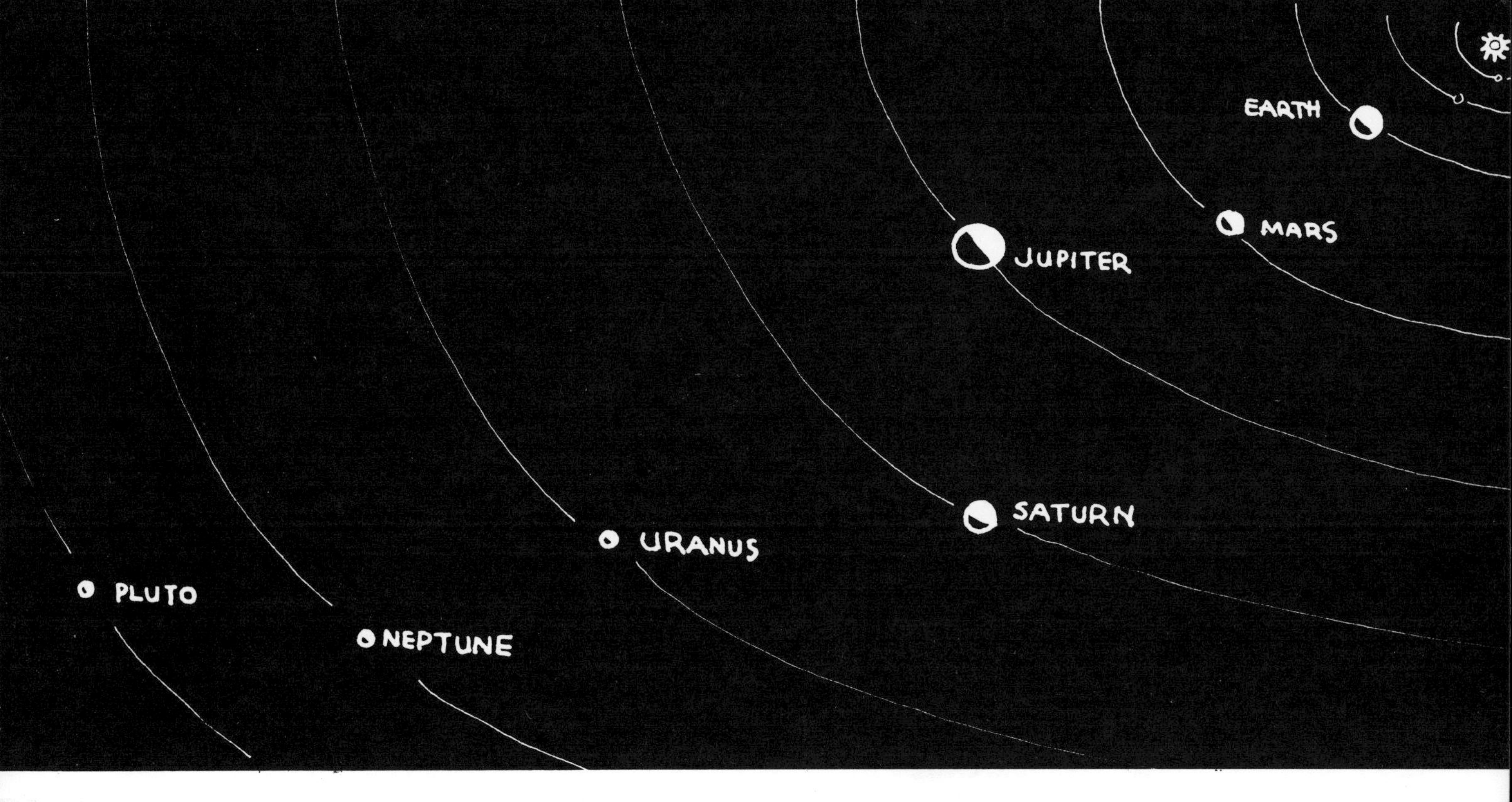

At midnight we can see the planets that are farther from the sun than we are. The planets Mars, Jupiter, Saturn, Uranus, Neptune, and Pluto can be seen in the middle of the night. That's because these planets can be on the opposite side of the earth from the sun.

We see Venus in our western sky in the evening after sunset. We see sunlight reflecting from the clouds that cover the planet. For about two months Venus seems to move away from the sun. Each night the planet appears a bit higher in our western sky, and sets a bit later. Then, as it moves along its curved orbit, it seems to us that Venus gets closer to the sun. It gets lower in the sky and sets earlier.

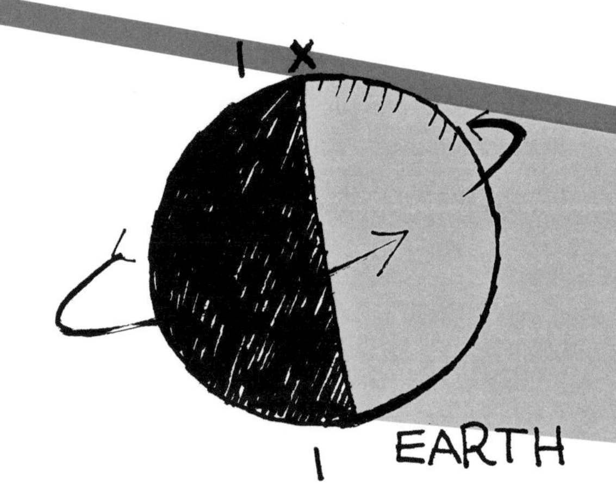

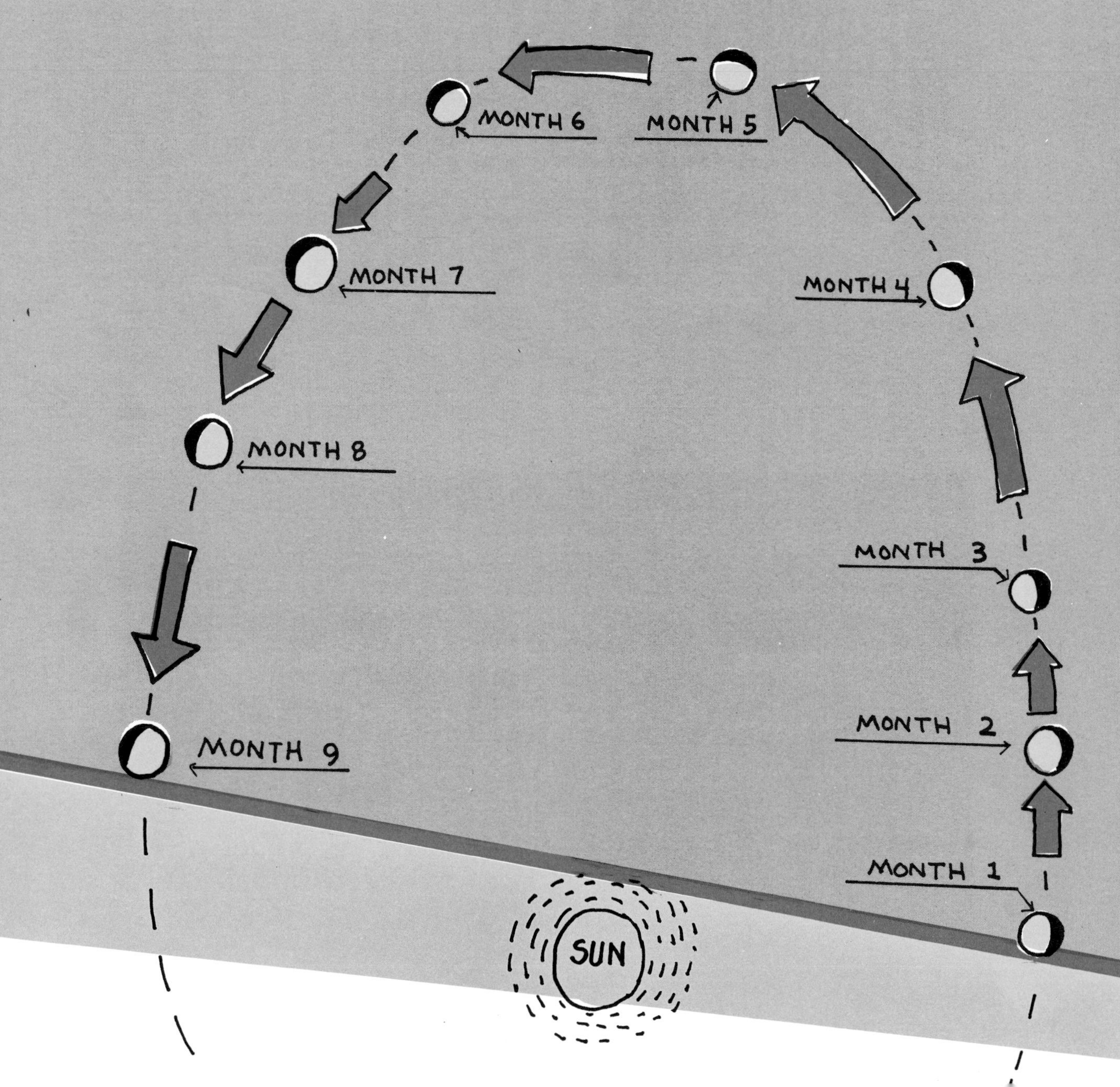
VENUS IN OUR WESTERN SKY
MONTH 6
MONTH 5
MONTH 7
MONTH 4
MONTH 8
MONTH 3
MONTH 9
MONTH 2
MONTH 1
SUN

For a few weeks Venus moves in between the earth and sun. We cannot see it at all because the lighted part of the planet is turned away from us. After a while the planet has moved to the other side of the sun. Then we begin to see it in our morning sky, just before sunrise. At first it can be seen just a few moments before sunrise. As the planet moves farther from the sun, we see it earlier and earlier. Then, as Venus continues to move along its orbit, the planet seems to move in toward the sun. The time of rising gets closer and closer to sunrise. Then we can no longer see it. That is because Venus is in our daytime sky. The next time we see the planet it will have become an evening star.

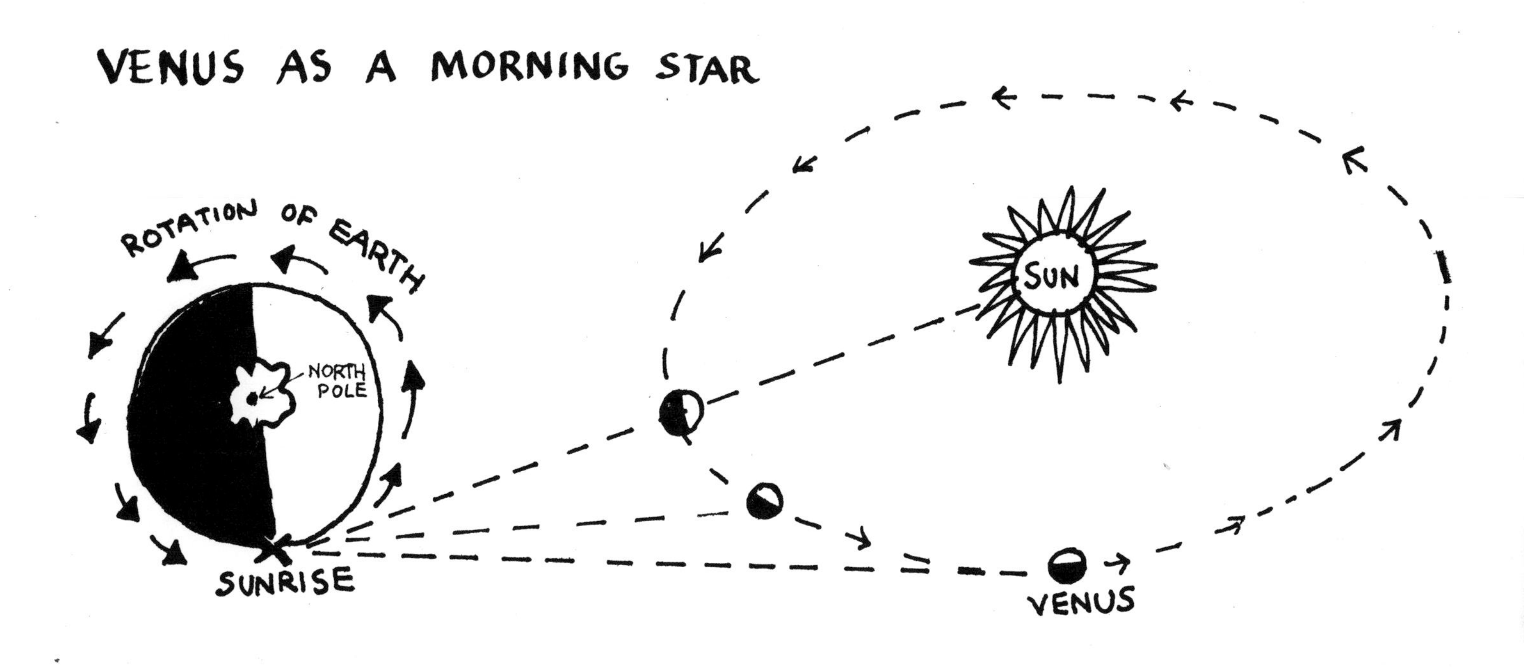
VENUS AS A MORNING STAR
ROTATION OF EARTH
NORTH POLE
SUN
SUNRISE
VENUS

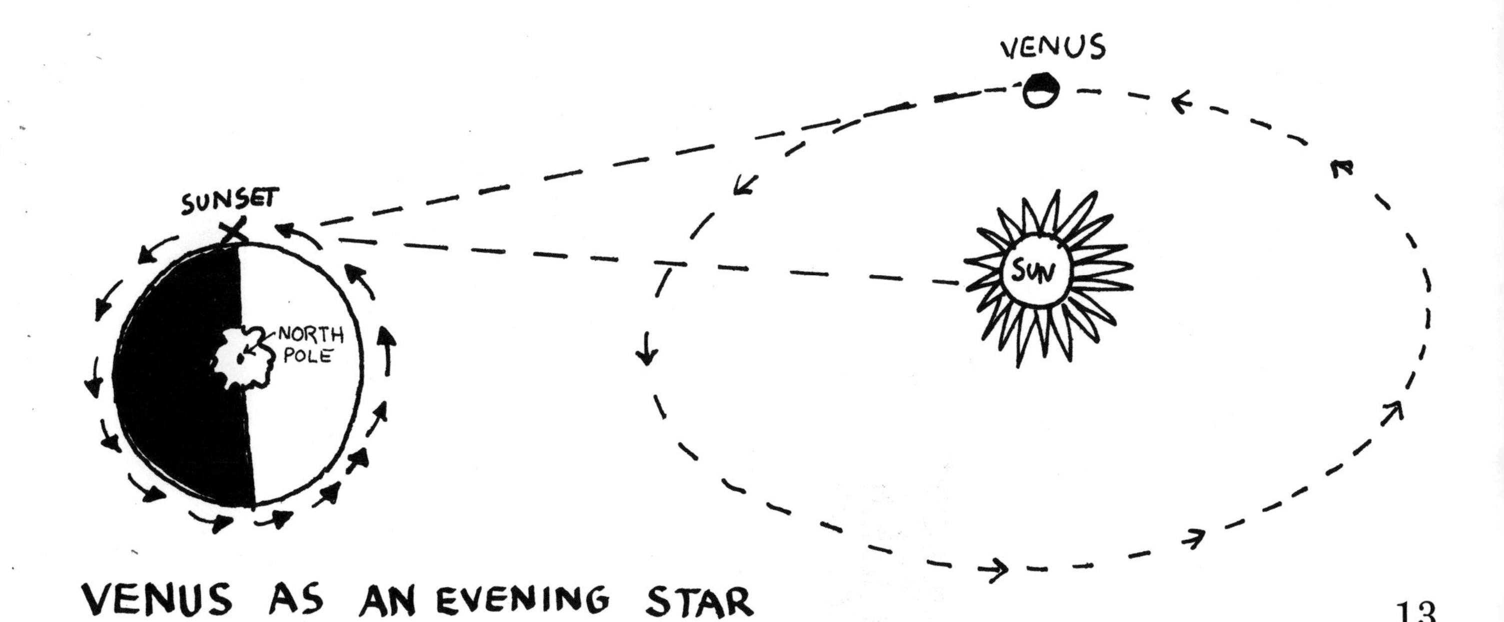
VENUS
SUNSET
NORTH POLE
SUN
VENUS AS AN EVENING STAR

You can see Venus very easily because the planet is brighter than the brightest star. When Venus can be seen in the evening (there is a list of dates on page 21), look for it toward the west—the part of the sky where the sun disappears below the horizon. Look there at sunset, and for an hour or so after sunset. You will see what appears to be a bright star. It will be Venus.

NAT KAPLAN

When Venus is a morning star, look toward the east—the part of the sky where the sun rises. Look there at sunrise, and an hour or two before sunrise. You will see what looks like a bright star. It will be Venus.

Venus is so bright anyone can see it easily. If you have a pair of binoculars or a telescope, you can see it even better. Make the glass steady by holding it against the corner of a building, or the back of a chair. Look at the planet carefully night after night (or morning after morning) and you will see something that Galileo, the famous Italian astronomer, discovered back in 1610. You will see the phases of Venus.

You know that the moon changes in appearance from full, to quarter, to crescent. These are the phases of the moon. One half of the round moon is always lighted by the sun. Sometimes we see all the lighted half. That's a full moon. Sometimes we see only a part of the lighted half. That's when we see a quarter moon or a crescent moon.

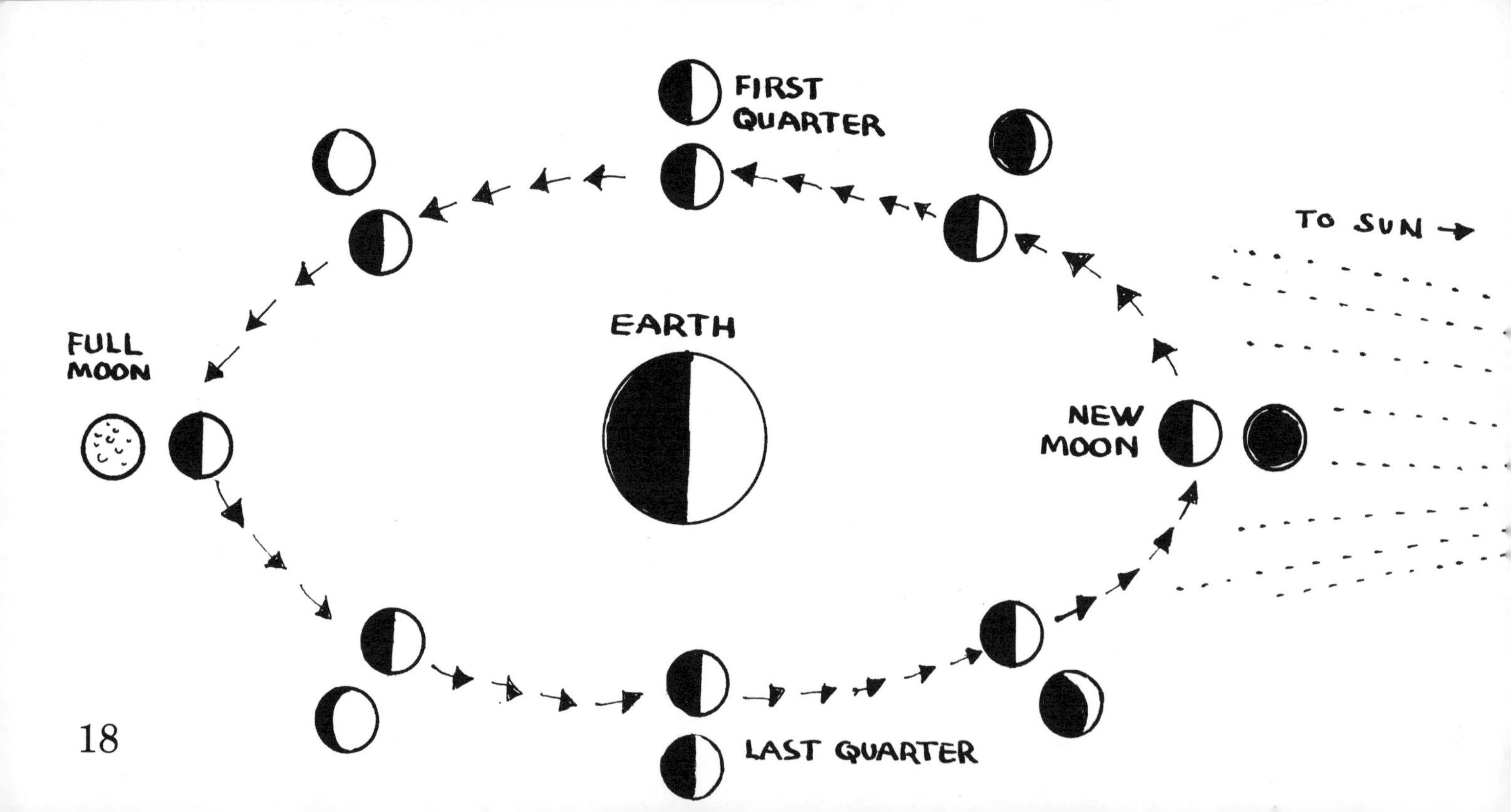

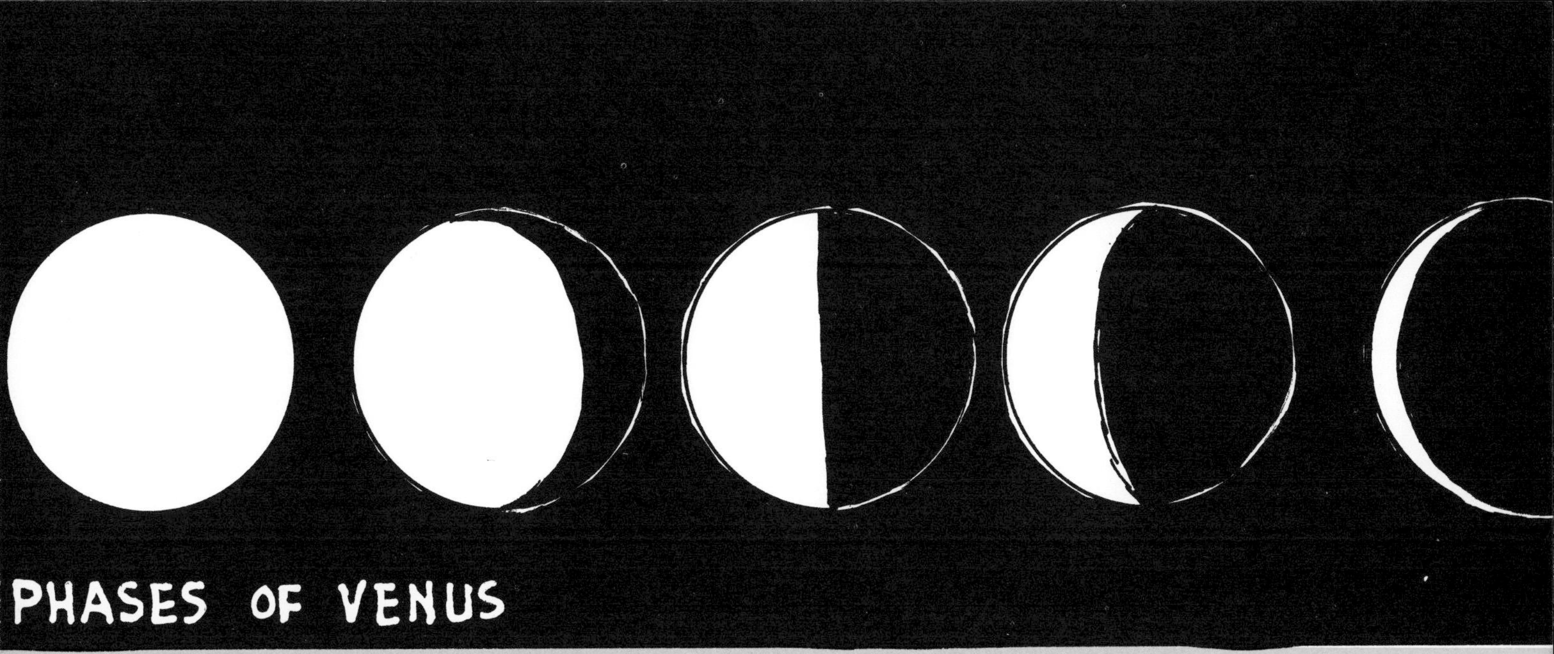

PHASES OF VENUS

The same is true of Venus. One half of the planet is always lighted by the sun. When we see all the lighted half of Venus, we see it full. When we see only part of the lighted half we see a quarter or a crescent. But you cannot see the phases of Venus unless you use good binoculars, or a telescope. With your eyes alone, the light blurs together.

The planet looks brightest when it is a large crescent close to the earth. Venus is full when it is farthest from the earth.

Here are the times for the next few years when Venus will be seen in the evening and in the morning. Be sure to look for it because, of all the planets in the whole solar system, Venus is the easiest one to find.

Year	*Evening – Brightest*		*Morning – Brightest*	
1969	Winter	February 25	Spring	May 15
1970	Fall	October 1		
1971			Winter	January 1
1972	Spring	May 8	Summer	July 27
1973	Fall	December 13		
1974			Spring	March 4
1975	Summer	July 18	Fall	September 7
1977	Winter	February 24	Spring	May 15
1978	Fall	September 29	Winter	December 18
1980	Spring	April 15	Summer	July 24

When you see Venus shining brilliantly in the evening sky, it is easy to understand why people down through the ages have wondered about this beautiful starlike world. Different people at various times through history have imagined it to be all manner of worlds.

The ancient Greeks and Romans associated the planet with mythology, considering it to be a goddess. The Romans called it Venus, meaning the goddess of beauty. The month of April was considered sacred to the goddess, and a day of the week—Friday—was dedicated to her. The Romans called this day "Dies Veneris." Our word "Friday" comes from the old Saxon word *Frigedaeg* (*Frigg* was the Saxons' name for Venus, and *daeg* means "day").

VENUS

Modern observers of Venus were more practical. They knew that the object was a planet, but still they could not see any of the surface. Therefore they allowed their imaginations to run wild. Around 1918, before astronomers had any information about the clouds around Venus, some people thought the clouds were made up of droplets of water and crystals of ice. They imagined that rain fell continuously from the clouds upon a surface that was very hot and wet. They believed that Venus might be a place where high forms of intelligent animal life could develop. Other astronomers believed that the surface of Venus was dry as dust.

Scientists have also passed the light from Venus through a spectroscope. This is an instrument that tells the scientist what materials are producing light or what materials might be reflecting it. Seen through a spectroscope, light from an electric bulb looks different from light produced by a fluorescent tube, for example, and light reflected by carbon dioxide looks different from light reflected by oxygen. Using a spectroscope, scientists discovered that there was a lot of carbon dioxide in the atmosphere of Venus. That discovery led to other theories about the nature of the surface of Venus. According to one of them, Venus was covered by water, and the water was bubbling with carbon dioxide, like soda water.

Such ideas were just theories. No one could prove that the conditions really existed or not, because no one could see the planet; no one could get a close-up view.

But on August 27, 1962, all the age-old theories were challenged. Early on that morning Mariner II was launched from Cape Kennedy in Florida. It was man's first effort to penetrate the space between the planets, and to probe the mysteries of another world.

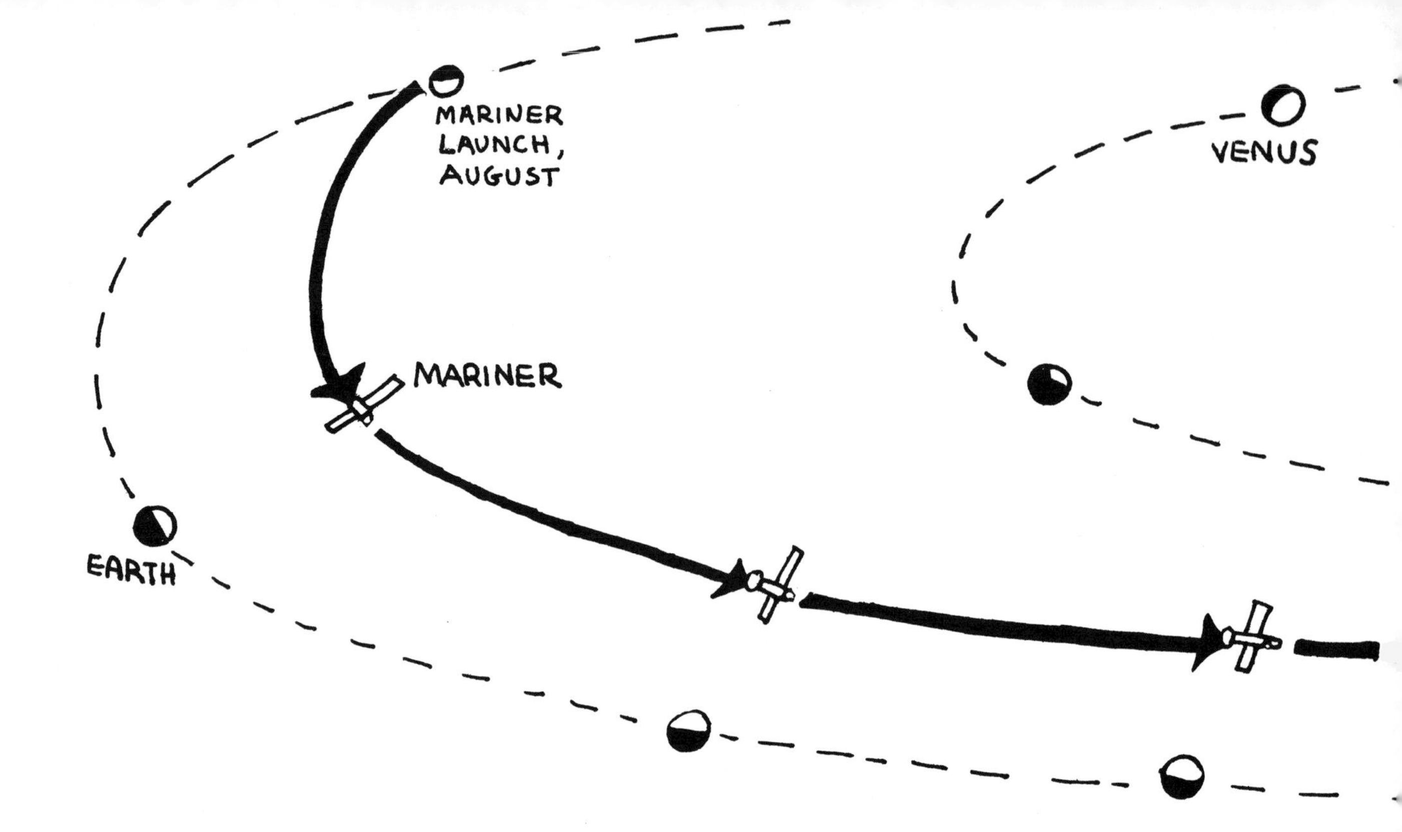

The rockets that sent Mariner II toward Venus from the earth put the spacecraft on a curving path that matched the curve of the orbit of Venus. The rockets detached from the spacecraft, and Mariner coasted toward Venus for four months. The path was about 180 million miles long.

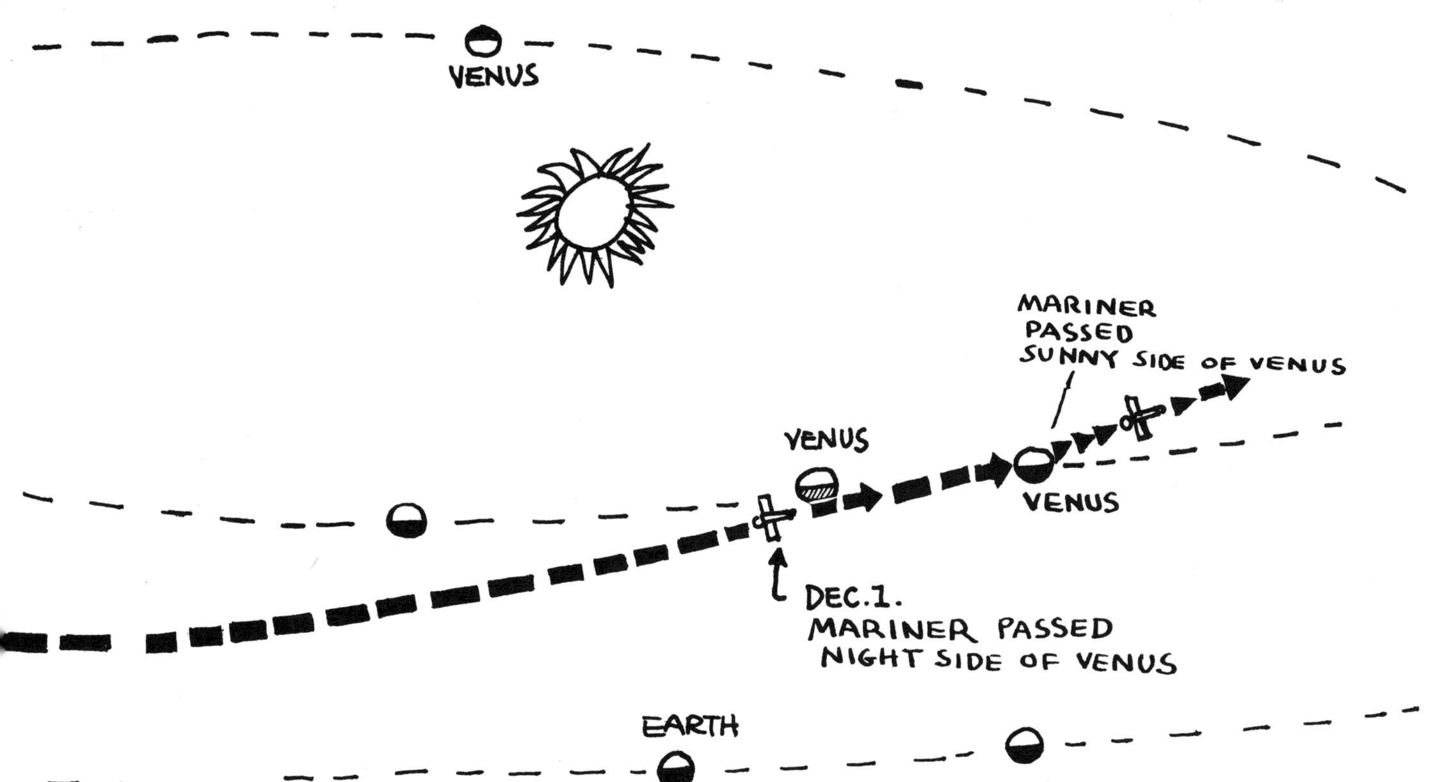

Mariner coasted from August to December. It was moving about 88,400 miles per hour as it moved close to the planet. When Mariner was a little more than 21,000 miles from Venus, instruments aboard were turned on by a command sent from the earth. Radios on the spacecraft sent to the earth the information gathered by the instruments.

Mariner is a spacecraft five feet across and ten feet high. Altogether it weighs about 500 pounds. Although we are no longer getting any information from Mariner, the craft is in an orbit around the sun right now, and will probably be there forever.

Mariner has a command antenna for receiving radio messages from the earth. It has another radio antenna that was used to send messages back to the earth.

One of the instruments aboard Mariner measured the magnetism of the planet. It showed that Venus has no magnetism at all. A compass would not work there.

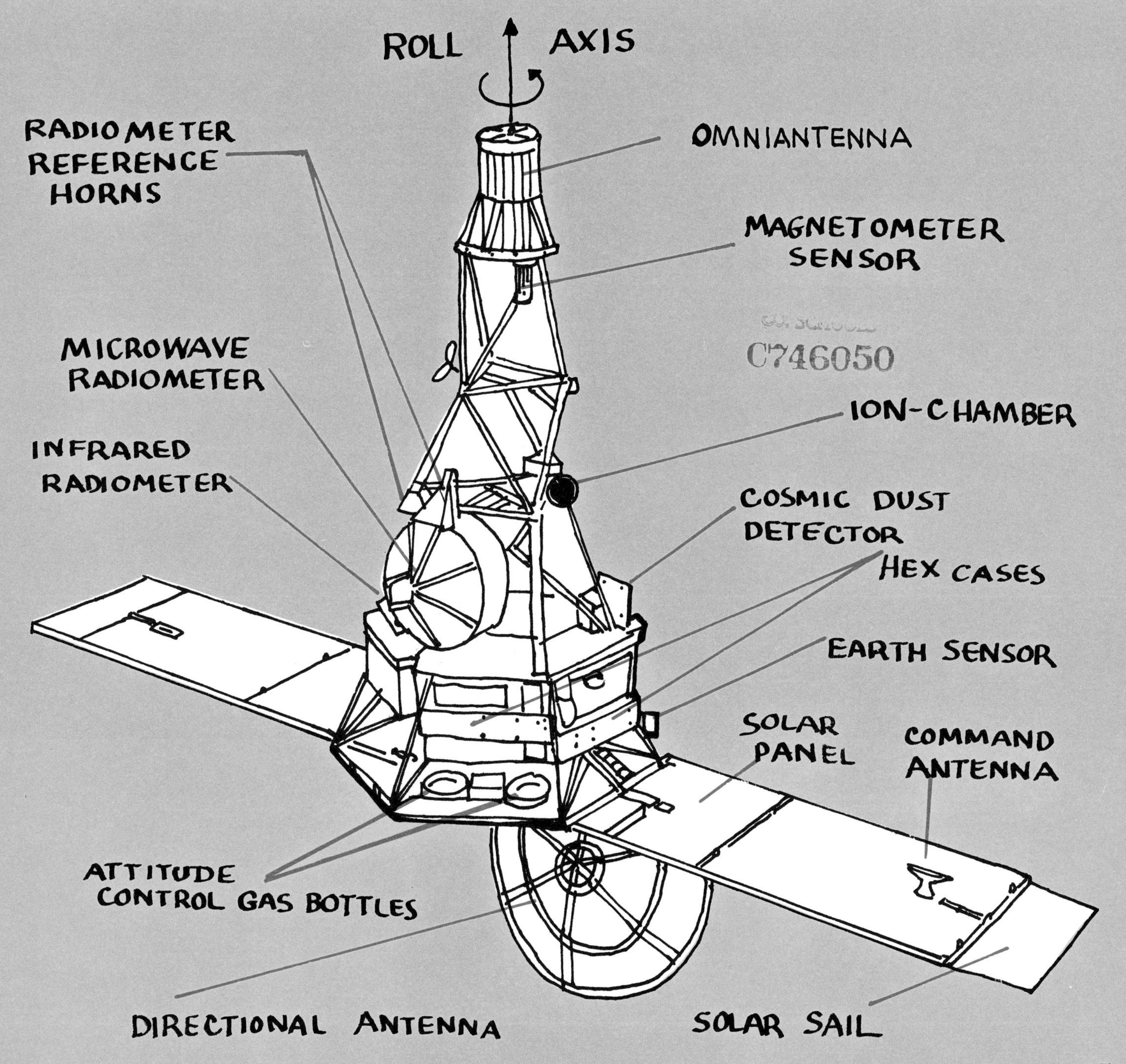
ROLL AXIS
RADIOMETER REFERENCE HORNS
OMNIANTENNA
MAGNETOMETER SENSOR
MICROWAVE RADIOMETER
ION-CHAMBER
INFRARED RADIOMETER
COSMIC DUST DETECTOR
HEX CASES
EARTH SENSOR
SOLAR PANEL
COMMAND ANTENNA
ATTITUDE CONTROL GAS BOTTLES
DIRECTIONAL ANTENNA
SOLAR SAIL

We do not know what causes earth's magnetism. It may be produced by solid layers in the earth that move slowly over hot, soft layers. Perhaps the reason Venus lacks magnetism is that the planet may be completely solid. Or Venus may lack magnetism because the planet spins around very slowly, much more slowly than earth does.

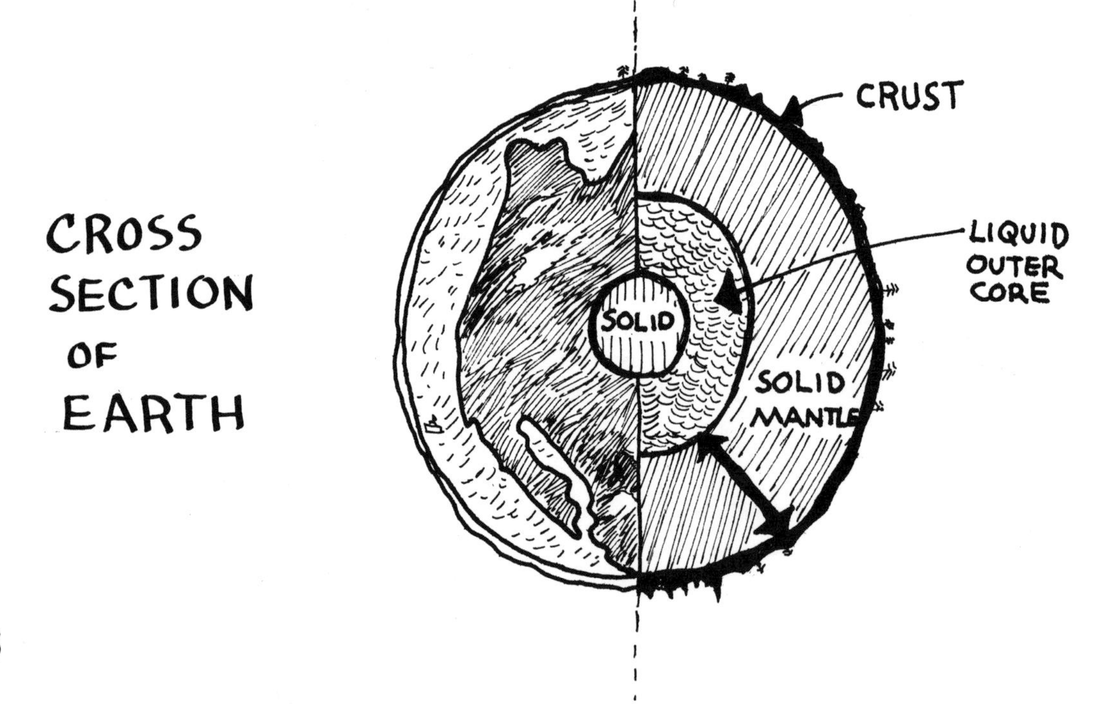

?
VENUS MAY
BE SOLID

Another instrument on the Mariner spacecraft measured the amount of dust in the space around Venus. We found out that there is much less dust there than there is around the earth.

Dust in space is called cosmic dust. It is also called meteoric dust. The particles are small, sometimes so small that a microscope is needed to see them.

The earth is inside so much dust that 1,200 tons of it fall to the surface every day. That's 10,000 times more than the amount of dust that falls on Venus.

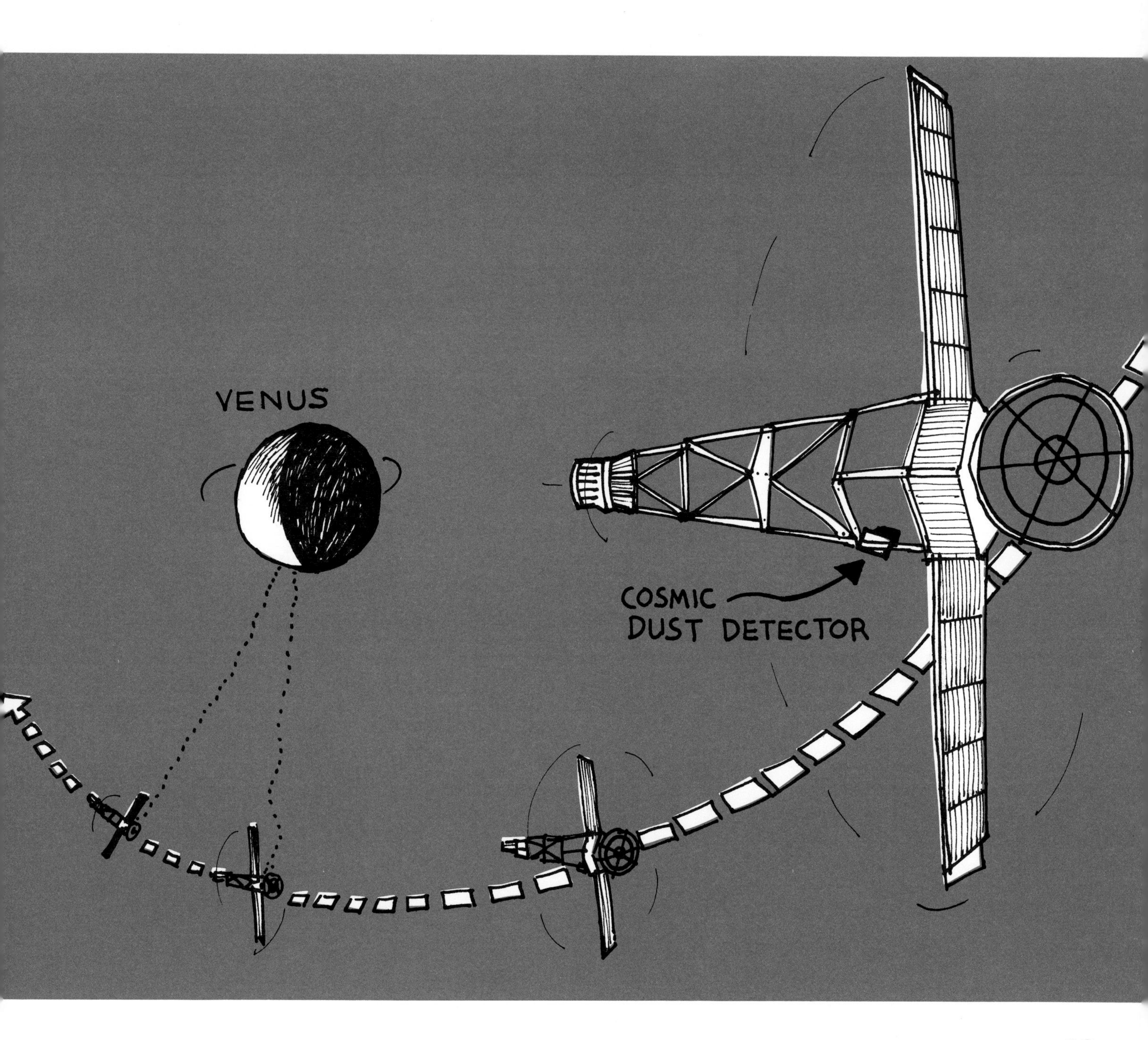
VENUS
COSMIC DUST DETECTOR

There are also other particles in the space around Venus. There are particles that are smaller than atoms. They are thrown out from the sun. These particles make up what is called the solar wind. They are ejected from the sun at high speed. They stream by Venus at speeds measured in millions of miles an hour.

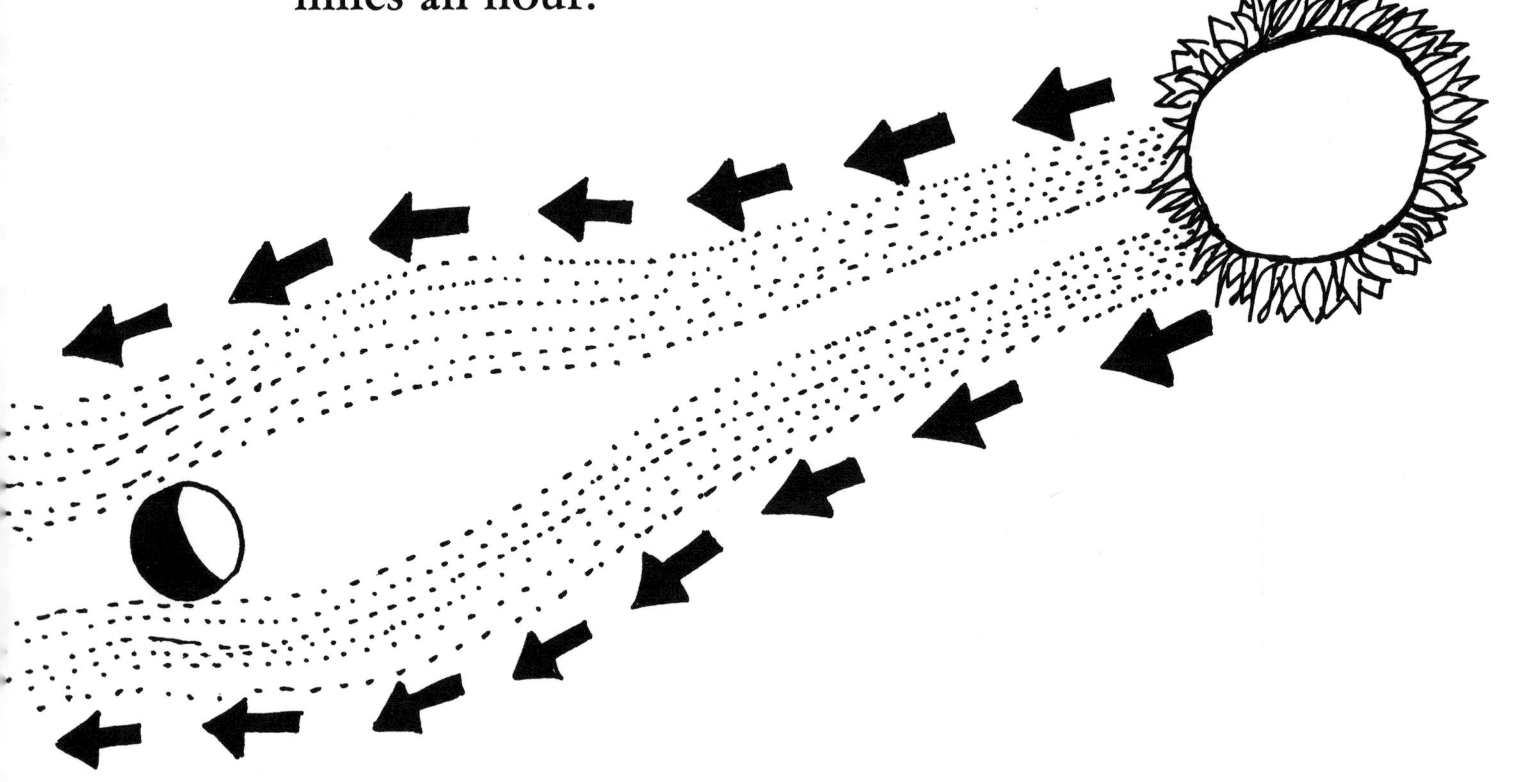

Mariner II also gave us a good idea of the temperature of Venus. At the surface it is around 625° Fahrenheit. That was a real surprise because no one thought the planet was that hot.

Mariner sent us information that was new and exciting. But we have been learning things about Venus in other ways too. For example, we have started to map Venus by radar. Radio waves are sent to Venus from the earth. They reflect from the planet, and come back to the earth. By carefully studying these reflected waves, scientists can tell something about the surface of Venus. The dark areas on this map made by radar are places where the Venus surface is rough. They may be mountains. Scientists are trying to map more of the surface.

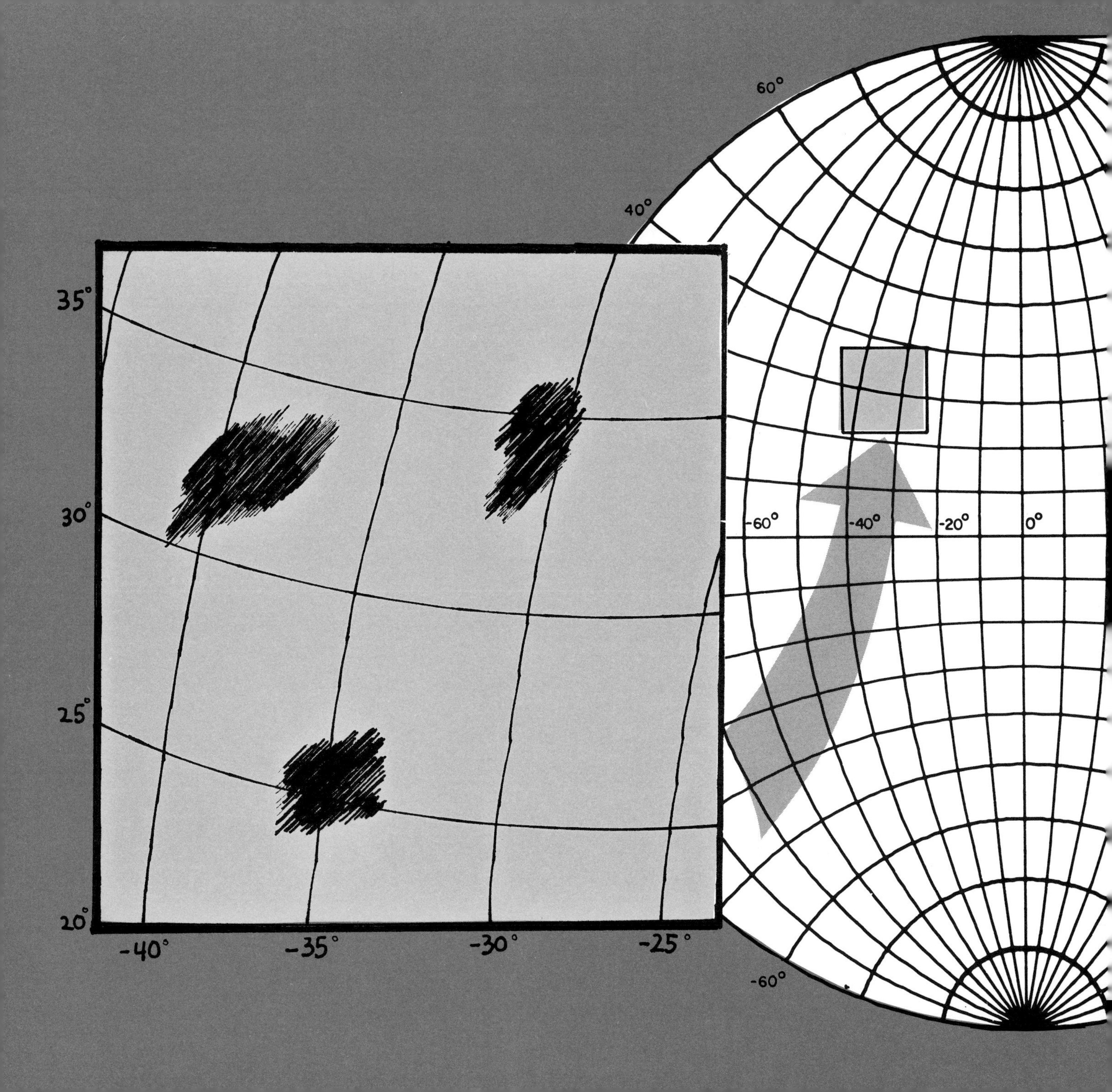

35°
30°
25°
20°
-40°
-35°
-30°
-25°
60°
40°
-60°
-40°
-20°
0°
-60°

We have other information about Venus. We know the mass of Venus is 10.729 septillion pounds. You can write the exact number like this:
10,729,408,500,000,000,000,000,000 pounds. That is less than the mass of the earth, which is 13.173 septillion pounds.

We use the term "mass" instead of "weight," because weight depends on where you are, and mass does not. If you weigh 100 pounds on the earth, you would weigh only 87 pounds on Venus. But your mass would not change.

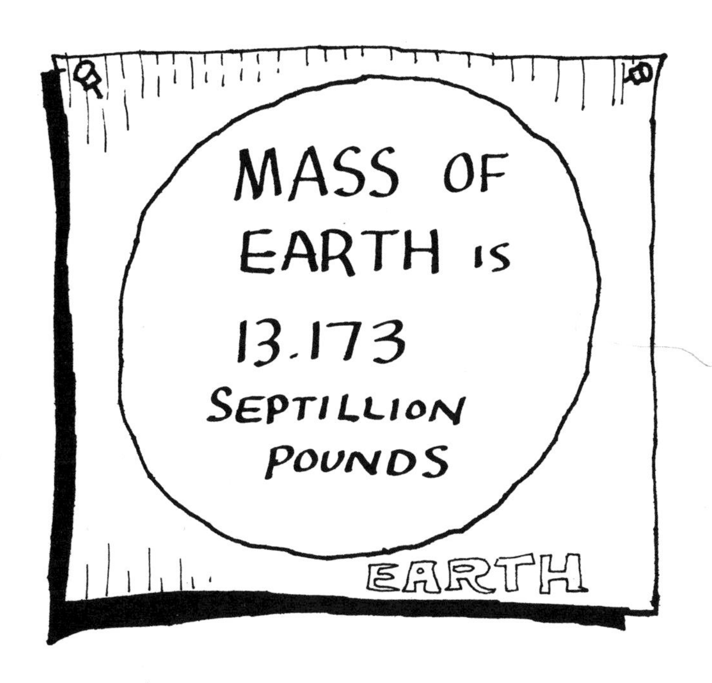

MASS OF VENUS IS

10,729,408,500,000,
000,000,000,000
POUNDS....

10.729
SEPTILLION
POUNDS

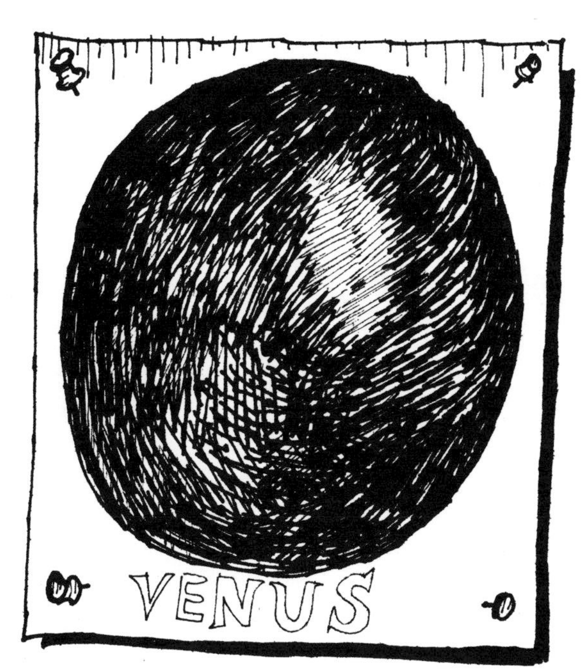

Venus is called earth's twin because the two planets are nearly the same size. The diameter of the earth is 7,927 miles, and the diameter of Venus is 7,700 miles. But Venus is different from the earth in many ways. It is most unusual because it rotates backward. Earth goes around the sun in a counter-clockwise direction as viewed from its north pole, opposite to the way clock hands move. It spins around (rotates) in the same direction.

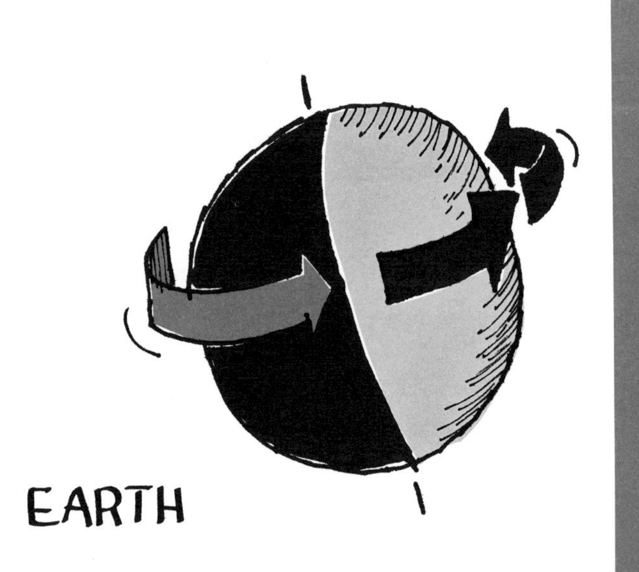

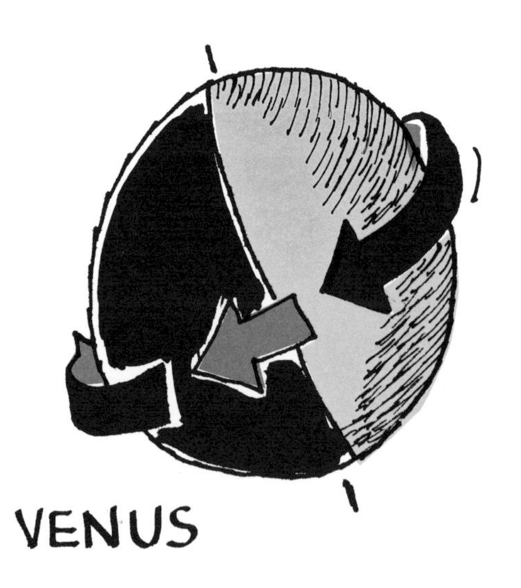

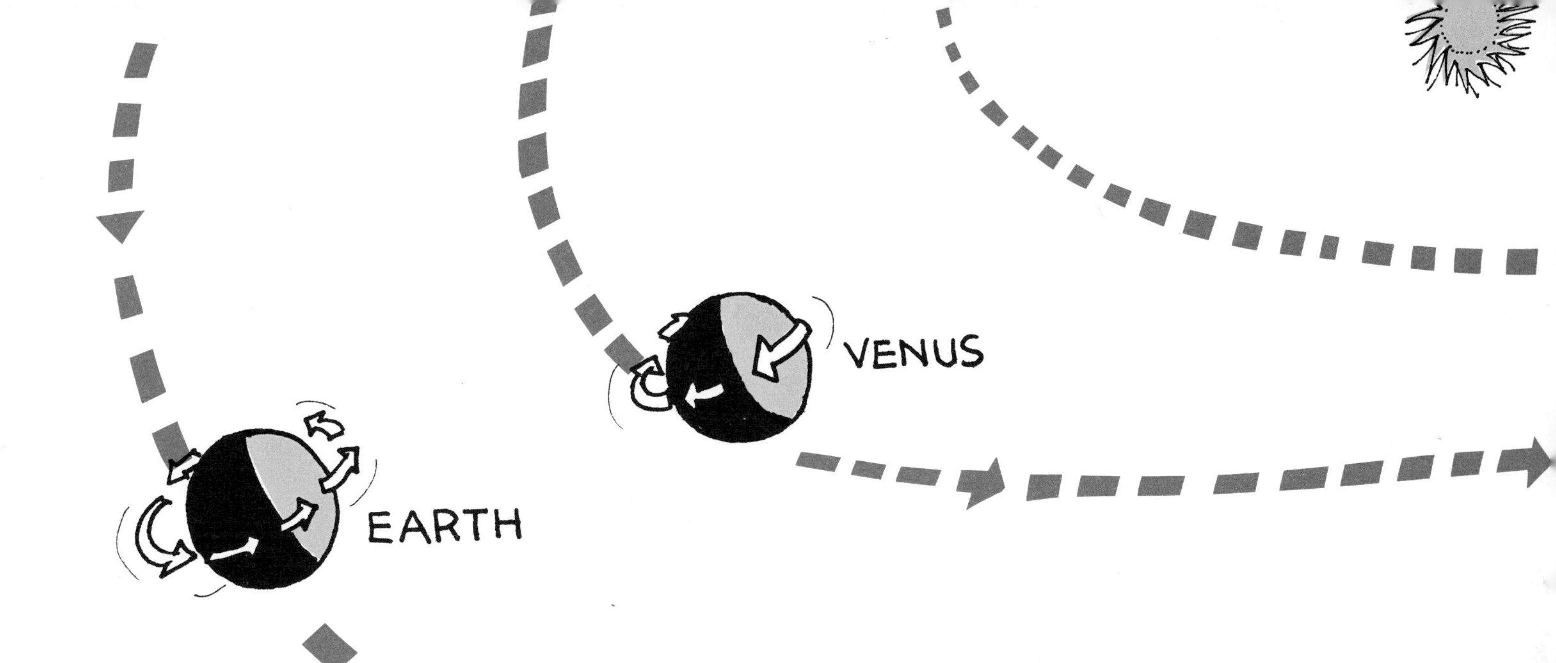

Venus goes around the sun in the same direction earth does. But Venus spins, or rotates, in the opposite direction. The planet rotates very slowly, however. It takes about 247 earth days for Venus to make a single rotation; while only 225 of our days are needed for Venus to make a complete journey around the sun.

On the earth a day is 24 hours long—that's how long it takes for a single rotation—and there are 365 days in one year (a complete journey around the sun). On Venus a day is 247 earth days long, and a year is made of 225 earth days. A day on Venus is longer than its year.

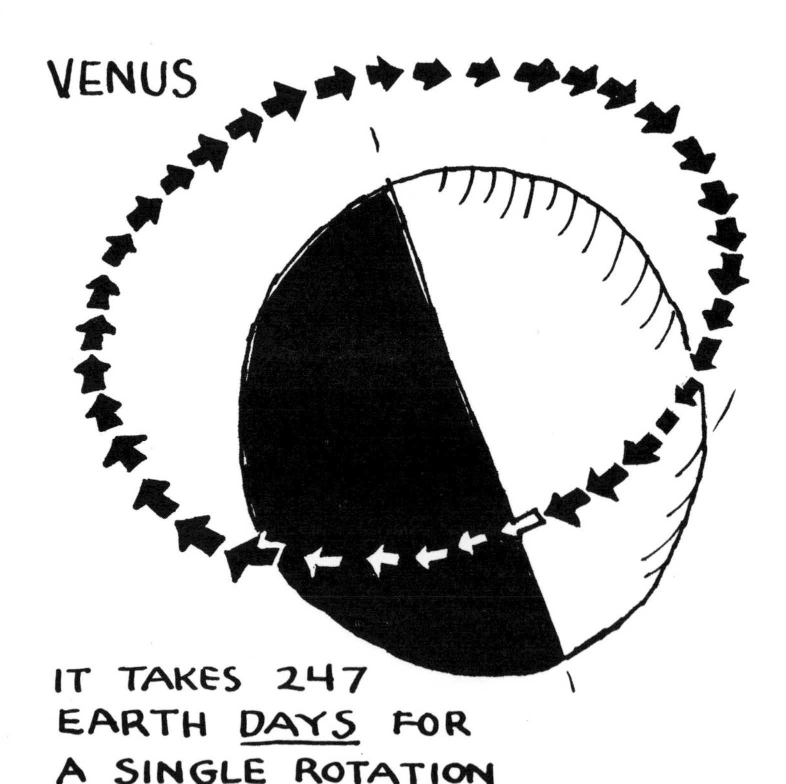

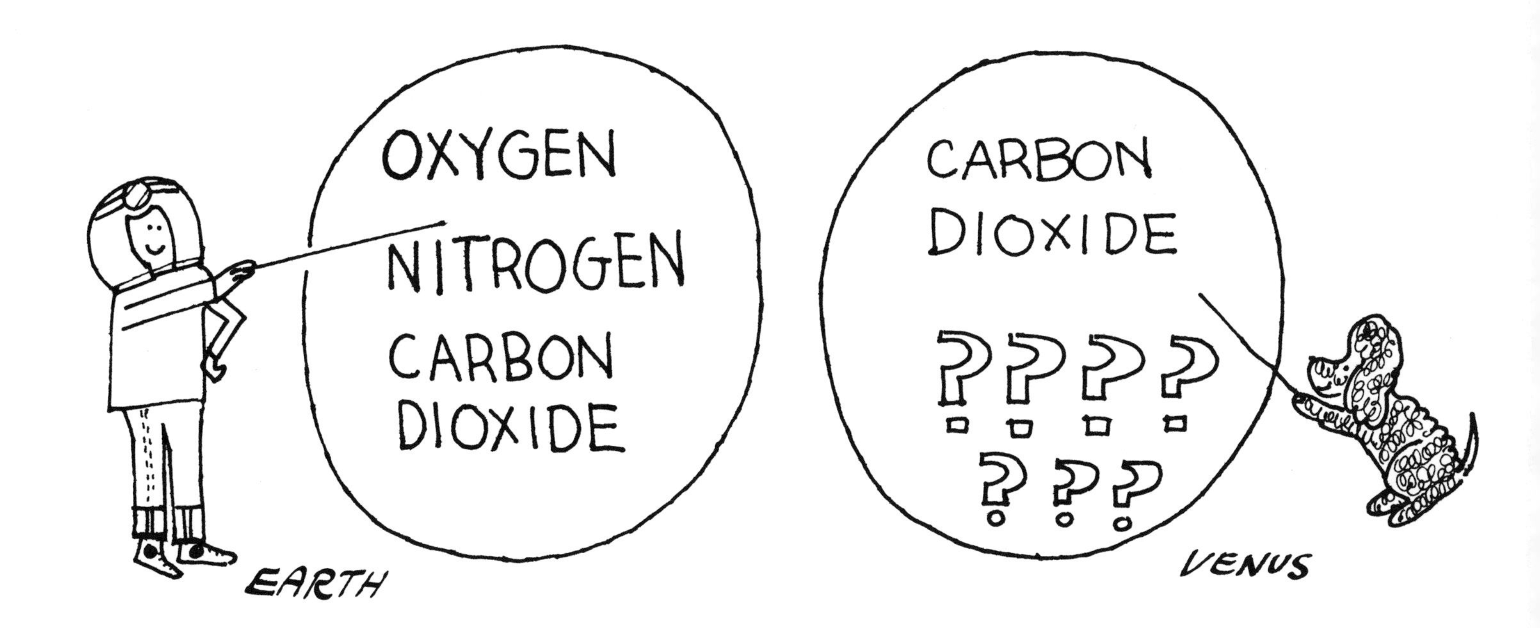

The atmosphere of Venus is unusual also. The air we breathe here on the earth is made up of oxygen, nitrogen, and a little carbon dioxide. We don't know exactly what the atmosphere of Venus is made up of. However, the spectroscope shows there is quite a lot of carbon dioxide in the atmosphere.

Carbon dioxide is the gas we give off when we breathe out. It is also a gas that plants need in order to grow. Some scientists believe that the carbon dioxide on Venus is frozen into crystals high in the clouds, and that's why we can't see through the atmosphere.

Recently scientists used a balloon to carry a spectroscope twenty miles above the earth. They pointed the spectroscope at Venus and discovered that there is some water vapor in the clouds of Venus. This could mean that there are also water droplets and ice crystals in various layers of the clouds. That would also make it impossible to see through the atmosphere.

In addition, the atmosphere of Venus may contain large amounts of nitrogen. We cannot say for sure because nitrogen cannot always be identified by a spectroscope. We have not been able to identify that gas on Venus. But it probably does exist there because it is found here on the earth and on those other planets that have atmospheres. Almost 80 percent of the air you breathe is nitrogen. Maybe 80 percent or more of the atmosphere of Venus is also nitrogen. We'll find out when new probes are sent to explore firsthand.

We are sure, though, that the atmosphere is very heavy. The pressure on the surface of Venus is twenty times more than the pressure of our atmosphere on the surface of the earth. Air pressure on you is about 14 pounds per square inch, or about a ton and a half on your whole body. On Venus the pressure on your body would be 20 to 30 tons.

Another mystery about Venus that has not been solved is the high temperature of the planet. Earth gets warm during the day. At night our atmosphere holds most of the heat. The air acts as a blanket. However, some of the heat escapes. Venus does not cool off at night. Probably that is because the atmosphere of Venus is so dense that heat cannot pass through.

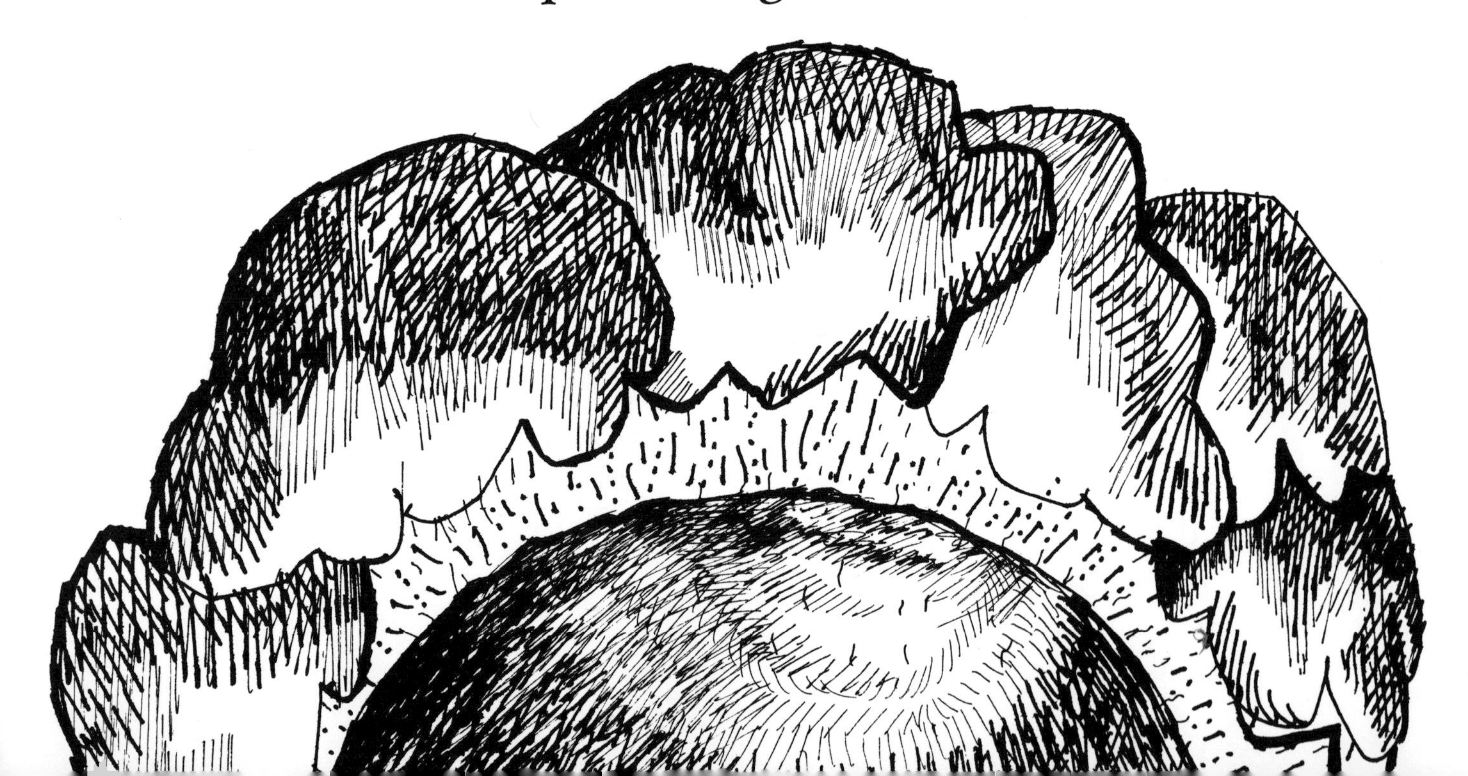

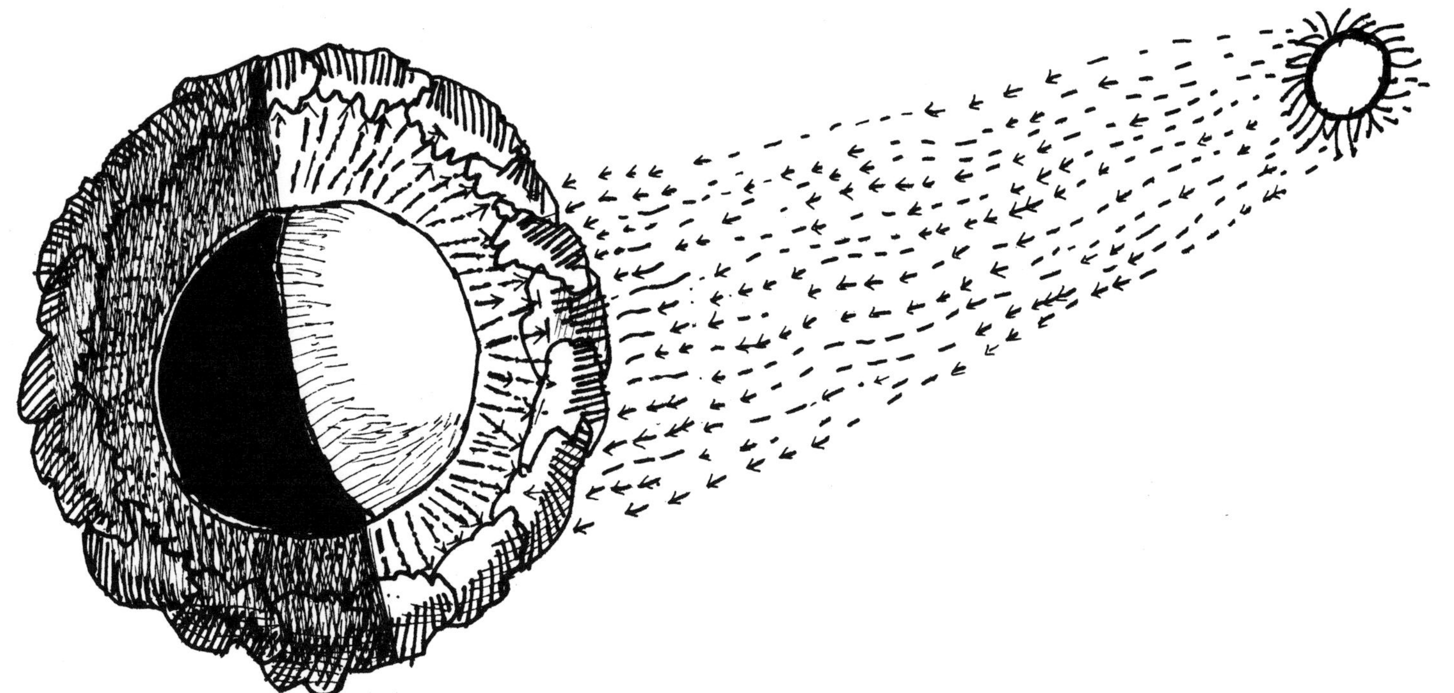

Venus gets its heat from the sun, just as we do. The energy that comes from the sun is called short-wave radiation. It can go through the clouds that surround Venus. When the energy reaches the surface of the planet, it changes to long-wave radiation. This radiation cannot go back through the clouds, and so it is trapped underneath them.

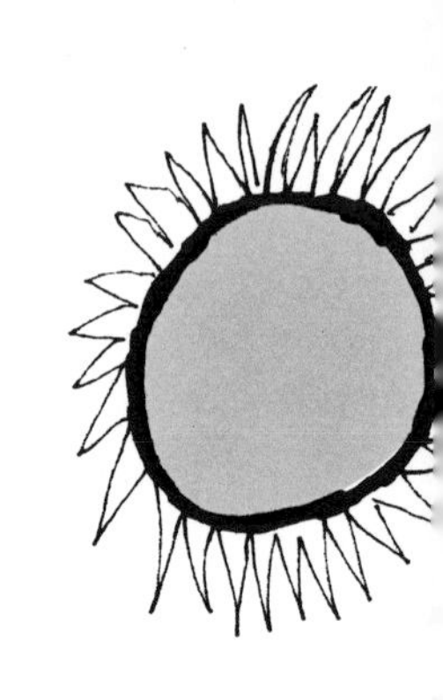

In winter, stand inside a room so that sunlight can fall upon you, and you'll see how this works. You feel warm in the sunlight. The energy from the sun, which is short-wave, passed through the glass window (the clouds), but the energy cannot pass out through the window because it becomes long-wave radiation when it falls upon you; short waves can pass through glass, but long waves cannot. This is called the greenhouse effect. In a greenhouse, solar energy (short-wave) goes through the glass. It changes to long-wave radiation when it falls upon the plants and soil. The energy is trapped inside the building and it keeps the plants warm.

On Venus, solar energy is trapped in the same way, probably. Some of the energy escapes, but enough is trapped to raise the temperature to about 625° F. High in the atmosphere of Venus the temperature is much lower. At some places it is as cold as 30° below zero, and farther from the surface the temperature is about 70° below zero.

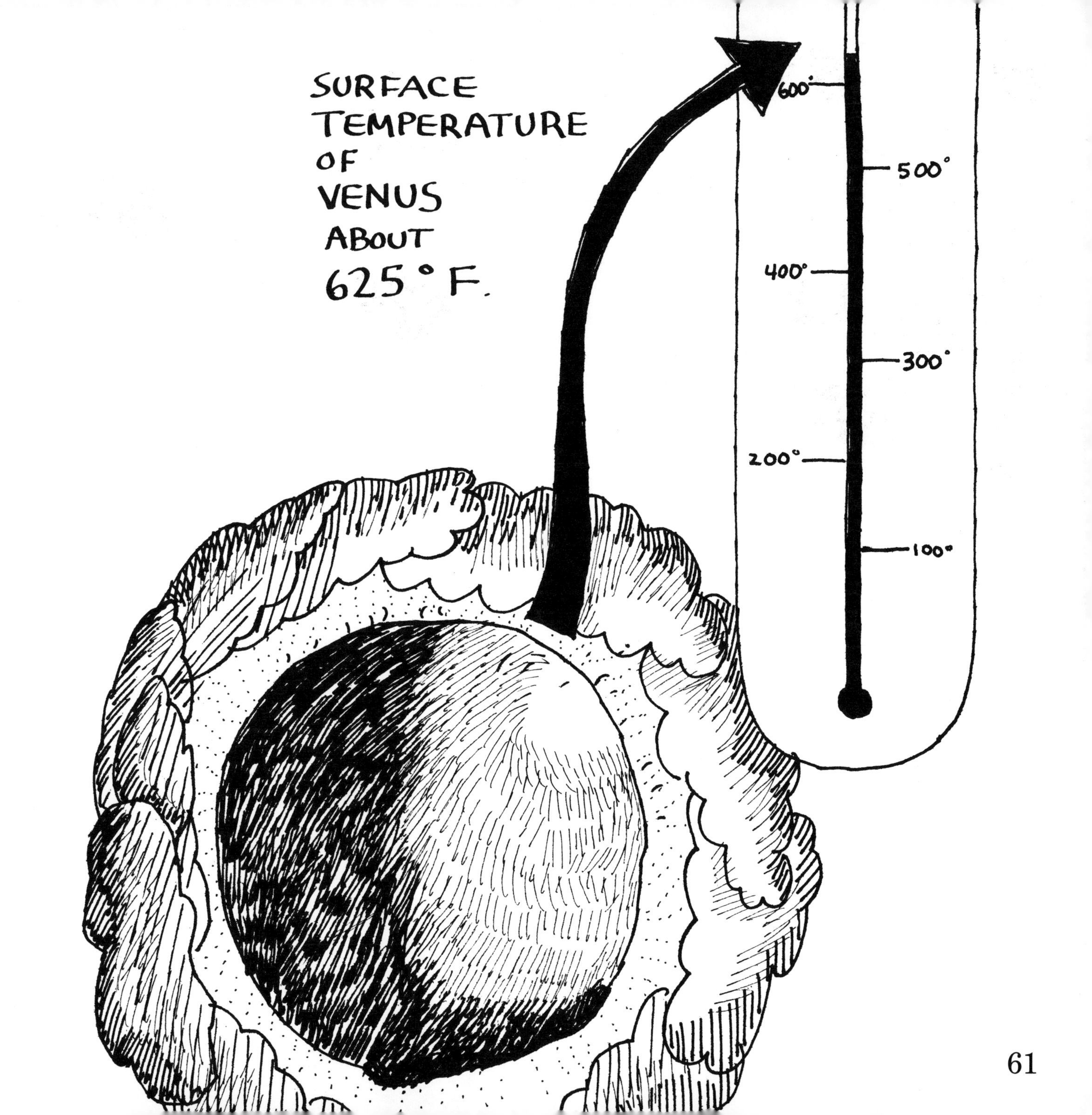
SURFACE
TEMPERATURE
OF
VENUS
ABOUT
625° F.
600°
500°
400°
300°
200°
100°

If the temperature all over the surface of Venus is 625° F. above zero, how could any form of life survive there? Maybe it couldn't. Maybe some day we'll find out whether or not there is any life on Venus, or ever has been. Some scientists tell us it might be possible for certain microorganisms to live there, even though it is very hot. Microorganisms—extremely small plants and ani-

mals—can adapt to all sorts of surroundings. Here on earth they can be found in hot springs, in snow fields, in the fuel tanks of jet airplanes where they live on kerosene, and in the cores that cool nuclear reactors where the radiation is so strong it would kill a man. So it may be that microorganisms could get used to living on Venus, even though the temperature is very high.

We shall try to find out whether there is life on Venus by landing instruments on the planet in much the same way that we have landed instruments on the surface of the moon. In 1967 Venera (Venus) 4 was landed on Venus by the Russians. We have not been told much about it, but apparently it was a sphere about three feet across. The sphere made a hard landing—that is, it crashed onto the planet. The radio inside it operated, however. It sent a radio signal for a short time and then became silent. We don't know why, but maybe the high temperature interfered with its operation. In 1969 two more Russian probes crashed onto the planet but they told us very little about it.

Other instruments will be landed on Venus in the future. Now that we know how hot it is there, the instruments and radios will be built to withstand high temperatures for quite a long time. One of the instruments will be a device to find out what the surface is made of. It will have a sticky "tongue" that will be stuck out. The tongue, actually a piece of sticky tape, will fall on the surface, picking up some of the dust of Venus. The tape will be pulled in and passed in front of a microscope. What the microscope "sees" will be sent by television back to the earth. In this way all of us here on the earth will get a look at a small part of the surface of Venus. We may discover that the planet is very much like the earth—or maybe the whole planet is a great, barren desert, as some people believe.

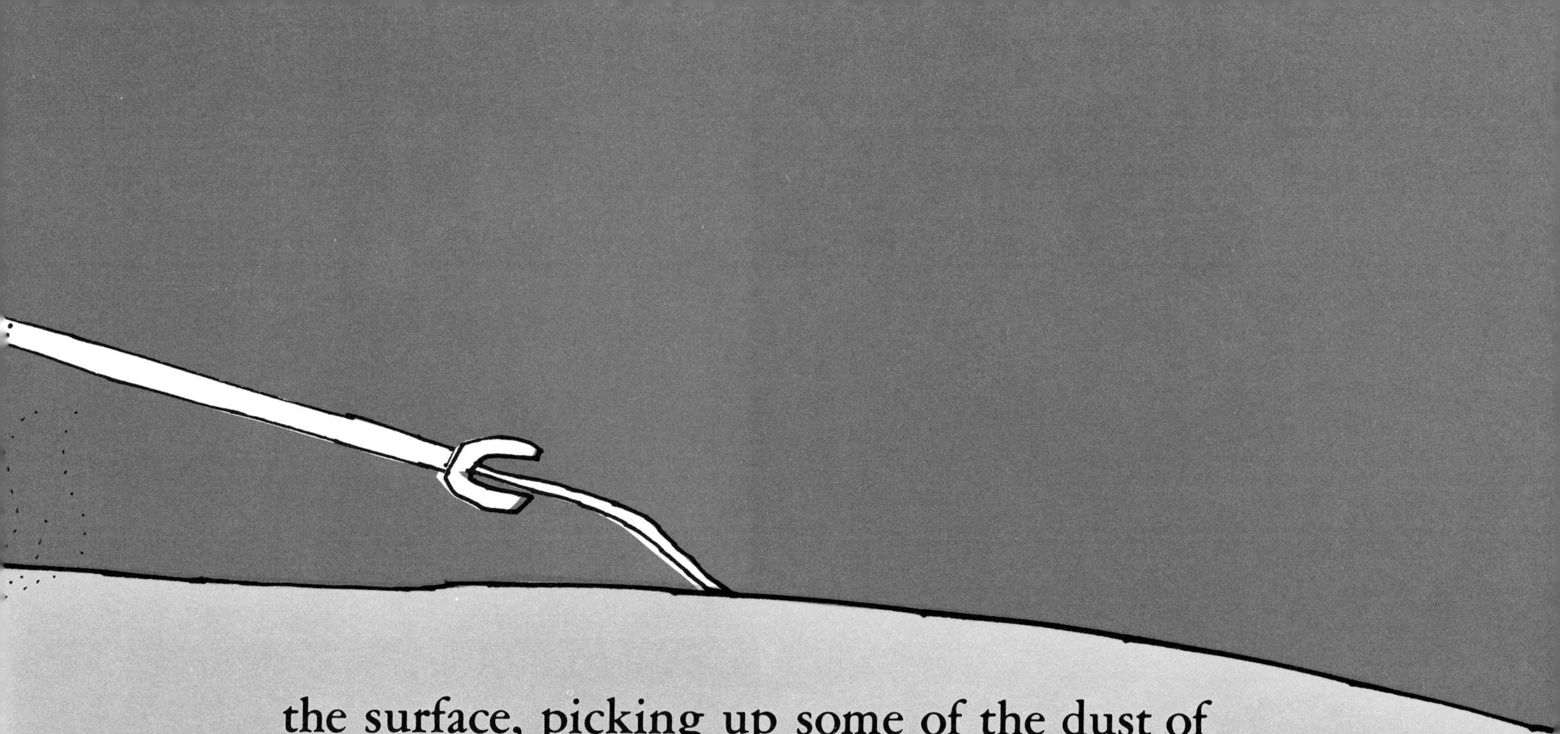

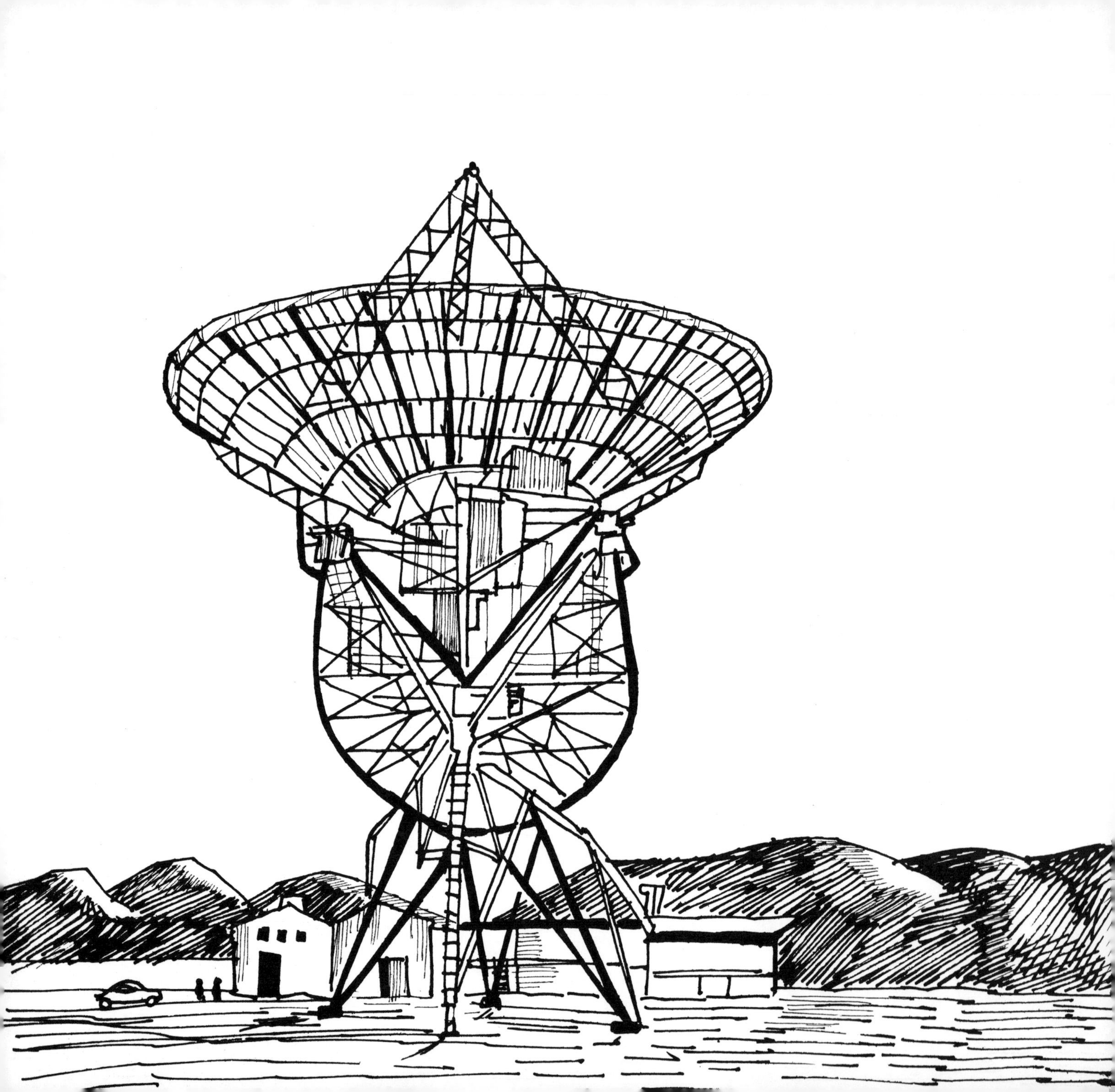

Mariner probes have passed close to Venus and "looked" at the planet carefully; astronomers have studied the planet with both optical and radio telescopes; the Venera probes have landed on the surface. We have gathered a lot of information about this brightest planet of all. But there are still many things we'd like to find out:

What is the surface really like? Is it rocky, or sandy, or covered with steaming swamps?

Does Venus have a magnetic field? So far we think not, but maybe better measuring instruments will tell a different story.

Is the planet solid all the way through?

What is the atmosphere made up of? And why can't we see through the atmosphere?

Is there any kind of life there?

If there is no life there now, has any kind of life ever existed there?

These are just a few of the questions we have about Venus, the second planet from the sun.

Venus is still a planet of mystery. We have solved some of its puzzles, and we can be sure that we'll find the solutions to more of them in the years ahead.

ABOUT THE AUTHOR

Dr. Franklyn M. Branley is well known as the author of many excellent science books for young people of all ages. He is also co-editor of the Let's-Read-and-Find-Out science books.

Dr. Branley is Astronomer and Chairman of the American Museum-Hayden Planetarium in New York City.

He holds degrees from New York University, Columbia University, and the State University of New York College at New Paltz. He lives with his family in Woodcliff Lake, New Jersey.

ABOUT THE ARTIST

Leonard Kessler is a writer and illustrator of children's books as well as a designer and painter. He became interested in children's books as a result of teaching art to young people in summer camps.

Mr. Kessler was born in Akron, Ohio, but he moved east to Pittsburgh at an early age. He was graduated from the Carnegie Institute of Technology with a degree in fine arts, painting, and design. Mr. Kessler enjoys playing the clarinet in his leisure time. He lives in New City, New York.